Kenneth Alvin Solomon
Anne J. Yatco

# Selected Topics in Biomechanical Injuries

Kenneth Alvin Solomon
Anne J. Yatco

# Selected Topics in Biomechanical Injuries

LAP LAMBERT Academic Publishing

**Imprint**

Any brand names and product names mentioned in this book are subject to trademark, brand or patent protection and are trademarks or registered trademarks of their respective holders. The use of brand names, product names, common names, trade names, product descriptions etc. even without a particular marking in this work is in no way to be construed to mean that such names may be regarded as unrestricted in respect of trademark and brand protection legislation and could thus be used by anyone.

Cover image: www.ingimage.com

Publisher:
LAP LAMBERT Academic Publishing
is a trademark of
Dodo Books Indian Ocean Ltd. and OmniScriptum S.R.L publishing group

120 High Road, East Finchley, London, N2 9ED, United Kingdom
Str. Armeneasca 28/1, office 1, Chisinau MD-2012, Republic of Moldova, Europe
Managing Directors: Ieva Konstantinova, Victoria Ursu
info@omniscriptum.com

Printed at: see last page
ISBN: 978-3-659-60864-3

# Table of Contents

# Chapter 1

## Introduction

Biomechanics is the study of the structure and function of biological systems such as humans, animals, plants, organs, and cells by means of the methods of mechanics (Alexander, 2005, p. R616-R619). The words "bio-mechanics" and "bio-mechanical" were coined by Nikolai Bernstein from the Ancient Greek words βίοσ (bios), meaning "life;" and μηχανική (mechanike), meaning "mechanics," to refer to the study of the mechanical principles of living organisms, particularly their movement and structure (Hatze, 1974, p. 189-190; *Oxford*, 2010).

In the practice of applying biomechanics to forensic science/accident reconstruction, the aim is to compare the expected movements of the body in response to the precipitating event (whether auto collision, Trip-and-Fall, Slip-and-Fall, misstep on a staircase, etc.), as well as the determined forces involved, to the common injury mechanisms associated with the specific injuries being claimed as a result of the precipitating event.

For example, a 55-year-old woman is involved in a collinear rear-end motor vehicle accident. Her vehicle is rear-ended with an estimated 5-mph change of velocity. The woman subsequently complains of shoulder pain and is eventually diagnosed with shoulder impingement syndrome. A biomechanist would first determine the expected movements of the body in response to the rear-end collision: the woman would move backwards (at 5 mph, relative to her vehicle) into her seatback (which would absorb a significant amount of her kinetic energy) and then rebound forward. Whether or not the woman held onto her steering wheel may be a contributing factor in whether or not the auto collision was a causative factor in her shoulder impingement syndrome; however, the low expected forces involved in the collision, as well as the lack of the common mechanisms associated with shoulder impingement, may lead the biomechanist to the conclusion that the subject motor vehicle accident was not the

2

proximate cause of the claimant's shoulder impingement syndrome. The ability to compare common injury mechanisms to the expected body movements during a precipitating event is important in evaluating the likelihood of a causal relationship between the two.

While an individual could indeed have been diagnosed with a shoulder impingement, the claimant's shoulder impingement may very well be a pre-existing condition, exacerbated by the subject accident. For example, a review of the claimant's medical records may reveal a history of shoulder complaints, and/or an MRI study (taken just a few months post-collision) that revealed evidence of degenerative changes and an anatomical configuration of the acromion that tends to be associated with impingement syndrome. These pieces of evidence would indicate that the claimant had a pre-existing condition at the time of the subject accident. Thus, a thorough understanding of the structures of the body and the common mechanisms which can produce injury would be just one part of the total analysis. Knowledge of the common mechanisms of various injuries, combined with a review of other evidence related to the injury claim, such as statements, photographs, medical records, etc., will aid the biomechanist and/or accident reconstructionist in determining the relationship, or lack thereof, between the precipitating event and the injury claim.

In this book, we consider the biomechanics of injury as it relates to accidents and injuries to the ankle, the knee, and the shoulder. In Chapter 2, we explore the anatomy and kinematics of the ankle, common ankle injury mechanisms, and integrative medicine treatments for ankle injuries. Chapter 3 is a comparison of two variations of the slip-and-fall accident, as well as an evaluation of the potential for ankle injury as a result of each type of slip-and-fall accident. Finally, in Chapter 4, we take a look at the shoulder, its kinematics, common mechanisms for shoulder injuries, and integrative medicine options for the treatment of shoulder injuries.

# References

Alexander, R. McNeill. (2005). Mechanics of animal movement. *Current Biology*, 15(16) , R616-R619.

Hatze, Herbert. (1974). The meaning of the term bio-mechanics. *Journal of Biomechanics*, 7, 189–190.

*Oxford English Dictionary*, Third Edition, November 2010.

# Chapter 2

## Ankle Injury Mechanisms and Integrative Medicine Therapies

(Originally published in the Winter 2012/Spring 2013 issue of the *Annals of Psychotherapy and Integrative Health*)

## Introduction

The purpose of this chapter is to show the relationship between the mechanism of an ankle injury (inversion, eversion, etc.) and the most likely result of the injury (sprain, fracture, etc.). While there are no absolute rules for positively associating each mechanism of injury with a specific type of injury, this chapter will provide some guidance for those attempting to prove or disprove the relationship between mechanism and injury type. Furthermore, this chapter will also illustrate common Integrative Medicine treatments for patients with ankle injuries, including acupuncture and homeopathic remedies. A preliminary examination of the anatomy of the ankle, as well as the kinematics that the structures of the ankle generate, will precede a discussion of types of injuries to the ankle. Following will be a discussion of the relationship between the type of injury and the mechanism of injury, from both an anatomical point of view and by example. Finally, we will discuss common Integrative Medicine treatments which patients with ankle injuries might undergo as part of their "whole body" treatment.

## Ankle Anatomy

The ankle is formed by the distal tibia, distal fibula, talus, and calcaneus (Figure 1). Superficially, the ankle's landmarks are the bony prominences on each side of the ankle, known as the medial and lateral malleoli, which are the rounded downward projections at the distal ends of the tibia and fibula, respectively (Figure 2).

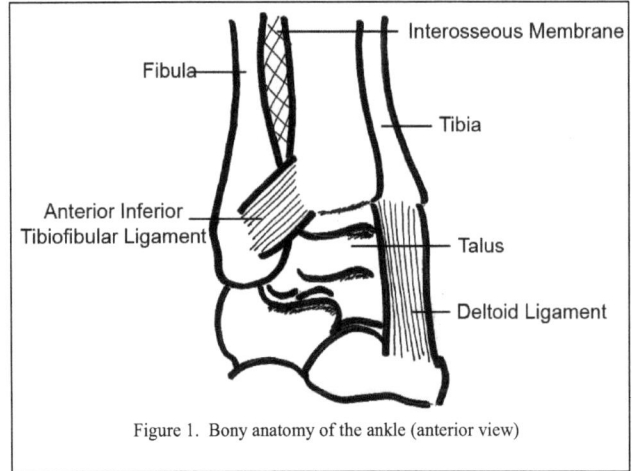

Figure 1. Bony anatomy of the ankle (anterior view)

The tibia, or shin bone, is the larger and stronger of the two bones of the leg below the knee and connects the knee to the ankle. The articular surface of the distal end of the tibia is also known as the plafond or pilon. The plafond articulates with the talus; together, they distribute weight bearing throughout the ankle (Small, 2009, p. 314).

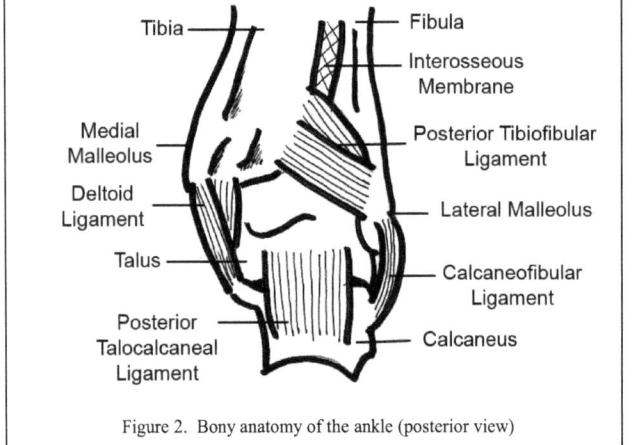

Figure 2. Bony anatomy of the ankle (posterior view)

The fibula, or calf bone, is located on the lateral side of the tibia. The fibula is connected to the tibia both proximally and distally. The lower extremity of the fibula projects below the tibia and forms the lateral portion of the talocrural joint. The fibula maintains ankle mortise stability during weight bearing.

The talus is the second largest of the tarsal bones. The superior, dome-shaped surface of the body of the talus is known as the trochlea. The calcaneus, or heel bone, meets the talus in two places: at the posterior and anterior talocalcaneal articulations. The ankle is surrounded by the articular capsule, which is attached to the borders of the articular surfaces of the malleoli proximally and to the distal articular surface of the talus distally (Norkus & Floyd, 2001, p. 69).

## Joints of the Ankle

The ankle is comprised of three joints: the talocrural joint, the subtalar joint, and the distal tibiofibular syndesmosis (Figures 3 and 4). The talocrural joint, also known as the tibiotalar joint or the mortise joint, is a uniaxial, modified-hinge joint formed by the medial malleolus of the tibia, the lateral malleolus of the fibula, and the talus. The tibial plafond articulates with the trochlea. The convex shape of the trochlea allows it to fit snugly into the concave plafond, which stabilizes the ankle mortise, the fork-like structure of the malleoli (Norkus et al., 2001, p. 68).

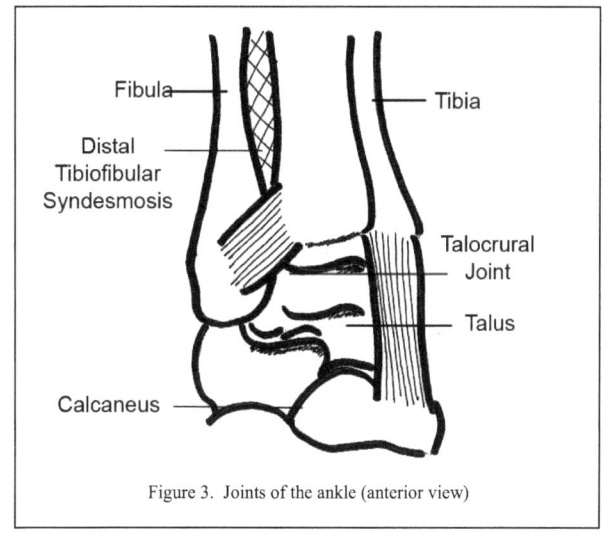

Figure 3. Joints of the ankle (anterior view)

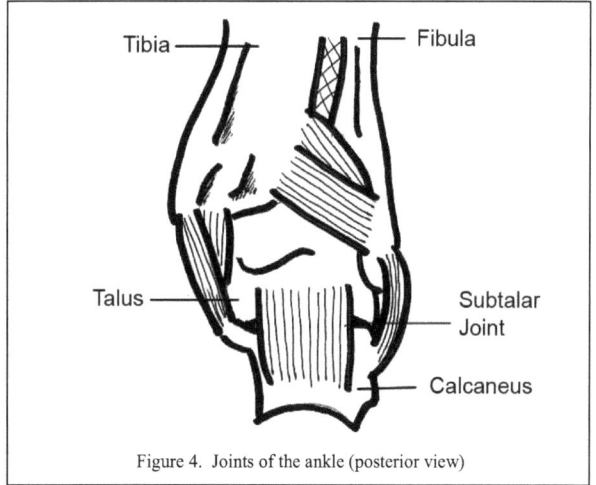

Figure 4. Joints of the ankle (posterior view)

This is important because the ankle bears more weight per unit area than any other joint in the body (Morrison & Kaminski, 2007, p. 135). The medial malleolus articulates with the medial aspect of the trochlea, and the lateral malleolus articulates with the lateral aspect of the trochlea. The talocrural joint allows for dorsiflexion and plantarflexion of the ankle. The normal range of motion (ROM) for the ankle joint is 30 degrees of dorsiflexion and 45 degrees of plantarflexion. Normal gait requires only 10 degrees of dorsiflexion and 20 degrees of plantarflexion (Small, 2009, p. 315). During dorsiflexion, the wider anterior portion of the talus occupies much of the mortise as it wedges itself between the medial and lateral malleoli; this is considered the safest ankle position due to the increased joint stability created by the increased contact of the articular surfaces of the talocrural joint.

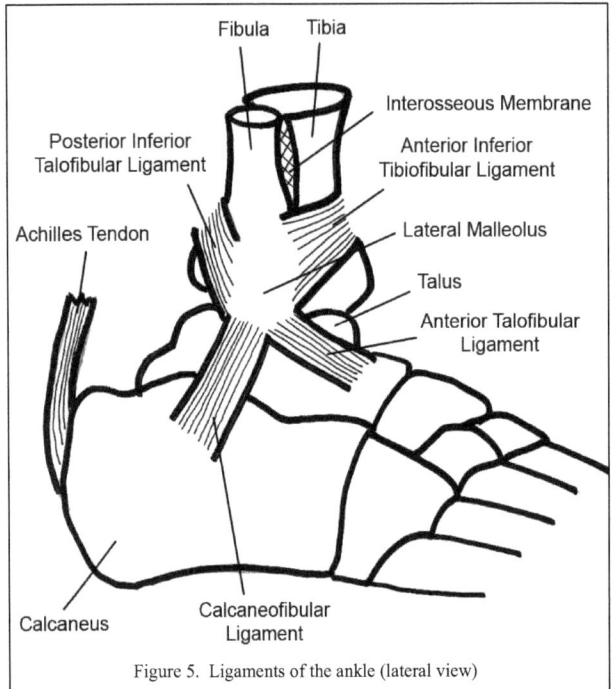

Figure 5. Ligaments of the ankle (lateral view)

The subtalar joint lies just inferior to the talocrural joint. The subtalar joint is a gliding joint, where the posterior aspect of the talus articulates with the superior aspect of the calcaneus. The anterior subtalar joint is formed by the head of the talus, the anterior-superior facets, the sustentaculum tali of the calcaneus, and the concave proximal surface of the tarsal navicular. The posterior subtalar joint is formed by the inferior posterior facet of the talus and the superior posterior facet of the calcaneus. The anterior and posterior subtalar joints behave like a single ball-and-socket joint. The subtalar joint averages a 42-degree upward tilt and a 23-degree medial angulation, allowing for inversion and eversion of the ankle (Fong, Chan, Yung, & Chan, 2009, p. 3).

The distal tibiofibular syndesmosis is a syndesmotic joint formed by the joining of the distal fibula and tibia by the anterior and posterior tibiofibular ligaments and the interosseous membrane (Molinari, Stolley, & Amendola, 2009). The distal tibiofibular syndesmosis allows for limited translation and rotation during dorsiflexion and plantarflexion, accommodating for the asymmetric talus (Fong et al., 2009, p. 3).

## Ligaments of the Ankle Joint

The talocrural joint is supported by the anterior talofibular ligament, the posterior talofibular ligament, the calcaneofibular ligament at the lateral aspect (Figure 5), and the deltoid ligament at the medial aspect of the ankle (Figure 6).

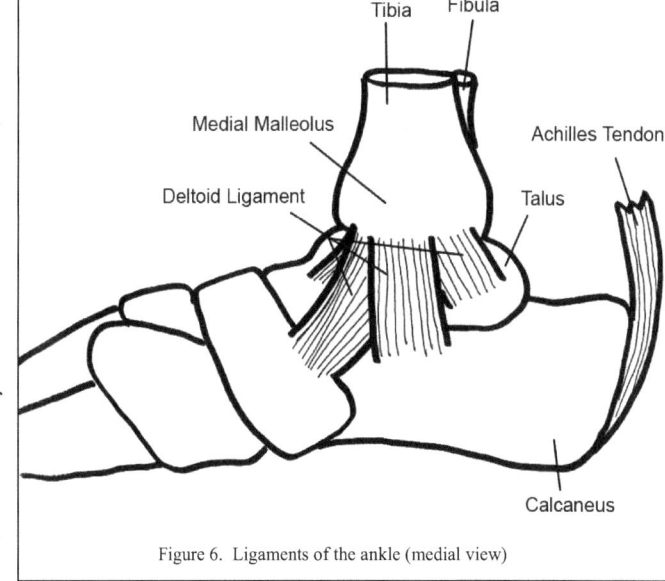

Figure 6. Ligaments of the ankle (medial view)

The anterior talofibular ligament originates from the anterior-inferior border of the fibula and inserts into the neck of the talus. It prevents anterior displacement and internal rotation of the talus during plantarflexion. It has the lowest ultimate load, approximately 138.9 N, and is the weakest of the lateral ligaments, making it the most susceptible to lateral ankle sprains (Fong et al., 2009, p. 3). The posterior talofibular ligament also connects the talus and the tibia and provides stability to the posterior aspect of the lateral ankle.

The calcaneofibular ligament connects the calcaneus and the lateral malleolus and limits ankle inversion. It is the strongest lateral ankle ligament. Injury to the calcaneofibular ligament occurs when the ankle is dorsiflexed and an inversion force is applied.

The deltoid ligament is a flat, triangularly shaped ligament found on the medial aspect of

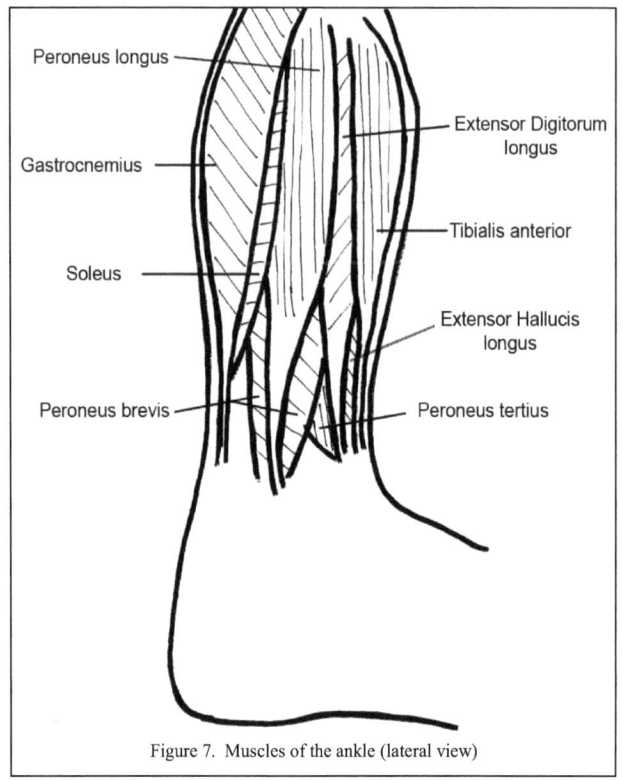

Figure 7. Muscles of the ankle (lateral view)

the ankle. The deltoid ligament has both a deep and a superficial portion. It is comprised of the anterior tibiotalar, the posterior tibiotalar, the tibiocalcaneal, and the tibionavicular bands. The deltoid ligament is considered to be the strongest ankle ligament. During plantarflexion, it prevents excessive eversion and resists talar external rotation. Injury to the deltoid ligament is uncommon and occurs due to excessive eversion (Norkus et al., 2001, p. 69; Small, 2009, p. 314).

The subtalar joint is supported by the deep ligaments, the peripheral ligaments, and the retinacula, which stabilize the subtalar joint and form a barrier between the anterior and posterior joint capsules. The three lateral ligaments also prevent excessive inversion and lateral talar tilt at the subtalar joint (Fong et al., 2009, p. 3; Norkus et al., 2001, p. 69).

The anterior and posterior tibiofibular ligaments, along with the interosseous

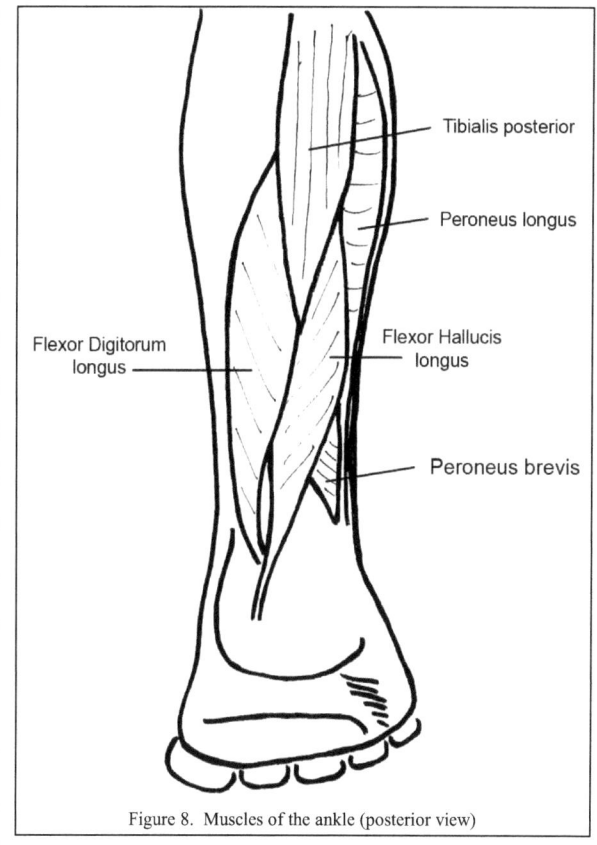

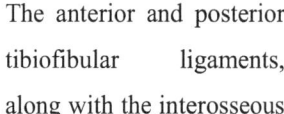

Flexor Digitorum longus

Tibialis posterior

Peroneus longus

Flexor Hallucis longus

Peroneus brevis

Figure 8. Muscles of the ankle (posterior view)

membrane, form a stable roof for the mortise of the talocrural joint and hold the tibia and fibula together (Figures 5 and 6). These syndesmotic ligaments resist axial and rotational forces against the ankle. The interosseous membrane also prevents posterolateral bowing of the fibula during weight bearing. If the tibia's articulation with the talus is shifted at all, the weight distribution on the talus can be altered, which may lead to early onset arthritis (Fong et al., 2009, p. 3; Molinari et al., 2009, p. 130; Norkus et al., 2001, p. 69; Small, 2009, p. 314).

## Muscular Control of Ankle Motion

The following muscles have tendons which pass behind the malleoli and act as ankle flexors: Peroneus longus, Peroneus brevis, Gastocnemius-Soleus complex, Flexor Hallucis longus, Flexor Digitorum longus, and Tibialis posterior. The following muscles have tendons which pass anterior to the malleoli and act as ankle dorsiflexors: Tibialis anterior, Extensor Hallucis Longus, Extensor Digitorum Longus, and Peroneus tertius (Martini & Bartholomew, 2000, p. 204-206; Patton, K.T., & Thibodeau, G.A., 2000, p. 228-229). (Please refer to Table #1 and Figures 7, 8, and 9.)

### Table #1.  Muscles of the Ankle

| Insertion of tendon | Muscle | Action |
|---|---|---|
| Behind the malleoli | Peroneus longus | Plantarflexion, eversion |
| | Peroneus brevis | Plantarflexion, eversion |
| | Gastrocnemius-Soleus complex | Plantarflexion |
| | Flexor Hallucis longus | Plantarflexion |
| | Flexor Digitorum longus | Plantarflexion |
| | Tibialis posterior | Plantarflexion, inversion |
| Anterior to the malleoli | Tibialis anterior | Dorsiflexion, inversion |
| | Extensor Hallucis longus | Dorsiflexion, inversion |
| | Extensor Digitorum longus | Dorsiflexion |
| | Peroneus tertius | Dorsiflexion, eversion |

## Tendons of the Ankle

The Achilles tendon is the most notable tendon of the ankle joint.  The Achilles tendon is a large tendon, running from the heel to the calf, shared by the

12

gastrocnemius and the soleus muscles, and it connects both muscles (as well as a third, vestigial muscle called the plantaris muscle) to the posterior calcaneus (Figures 5 and 6). The attachment of the gastrocnemius and soleus muscles to the calcaneus allows for plantarflexion of the ankle. The Achilles tendon is a strong, nonelastic, fibrous tissue that can absorb large forces associated with running, upwards of six to eight times the body's weight (Dubin, 2005, p. 39).

## Bursae of the Ankle

A bursa is a sac containing a viscid fluid that helps to reduce friction between moving parts. Bursae are usually found over bony prominences and beneath tendons. The bursae with the most clinical significance are the retrocalcaneal bursa, located between the Achilles tendon insertion site and the calcaneus, and the retroachilles bursa, located between the Achilles tendon and the skin (Aldridge, 2004, p. 334). Injury to the

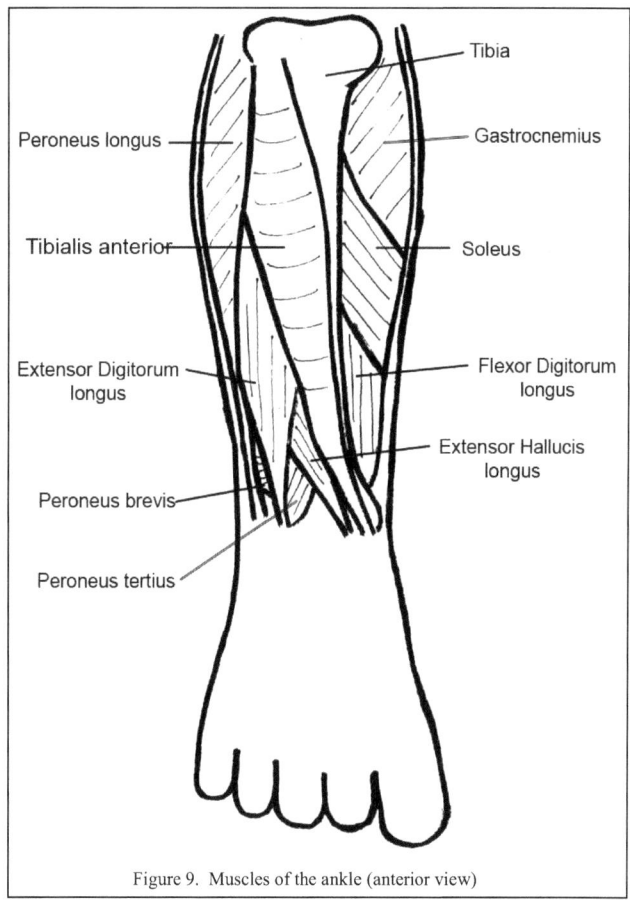

Figure 9. Muscles of the ankle (anterior view)

13

bursae of the ankle is common, and may be concurrent with another injury to the ankle.

## Table #2. Mechanisms & Examples of Ankle Injuries

| Injury | Mechanism of Injury | Examples of How Injury Occurs |
|---|---|---|
| Lateral Ankle Sprain | inversion and plantarflexion | "twisted ankle" |
| Medial Ankle Sprain | eversion | landing on the inside of the foot after jumping |
| High (Syndesmotic) Ankle Sprain | external rotation and hyperdorsiflexion | a blow to the lateral leg while the foot is planted, falling forward onto a planted foot |
| Unimalleolar Fracture | supination-adduction, supination-external rotation, pronation-abduction, and pronation-external rotation | falling onto the top of the foot |
| Bimalleolar Fracture | supination-adduction, supination-external rotation, pronation-abduction, and pronation-external rotation | falling onto the outside of the foot while the foot is planted (pronation-abduction) |
| Trimalleolar Fracture | supination-adduction, supination-external rotation, pronation-abduction, and pronation-external rotation | excessive rotational force while the foot is planted |
| Ankle Arthritis | posttraumatic degenerative joint disease; may also be osteoarthritic, rheumatoid, or septic in nature | prior trauma, age |
| Tendonitis | tight gastrocnemius and soleus muscles, overpronation of the subtalar joint, degeneration | overuse of the calf muscles, abnormal stress on the ankle, age |
| Bursitis | irritation of the area of the Achilles tendon insertion at the posterior calcaneus | ill-fitting footwear |

## Common Ankle Injuries

Common ankle injuries include, but are not limited to, ankle sprains, ankle fracture, arthritis, tendonitis, and bursitis. Please refer to Table #2: Mechanisms & Examples of Ankle Injuries.

## Ankle Sprains

Ankle sprains are the most common type of ankle injury. Sprains occur as a result of the stretching or tearing of any of the ligaments surrounding the joints. Sprains are typically classified as a grade I, II, or III sprain. A grade I sprain is considered to be a mild sprain, involving stretching or inflammation of a ligament; a grade III sprain, on the other hand, is a complete tear of a ligament. At this stage, the patient's ankle will suffer a complete loss of function and motion, as well as mechanical instability (Ardizzone & Valmassy, 2005, p. 65; Wolfe, Uhl, Mattacola, & McCluskey, 2001, p. 93). The three types of ankle sprains are lateral, medial, and high (syndesmotic).

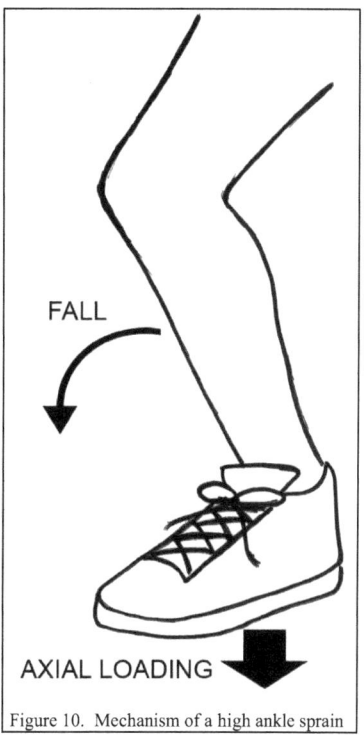

Figure 10. Mechanism of a high ankle sprain

Lateral ankle sprains are the most common. The anatomy and biomechanics of the ankle puts the lateral ankle at the highest risk to sustain inversion injuries; in fact, 85% of ankle sprains are caused by an inversion mechanism (Morrison et al., 2007, p. 135). Lateral sprains often occur from an inversion (or supination) force applied to a foot in plantarflexion. For example, the common "twisted ankle," in which the foot rolls inward and the patient lands on the outside of the foot, is a lateral sprain. The anterior talofibular ligament is injured first and, if the force is great enough, the calcaneofibular ligament follows (Small, 2009, p. 316).

Medial ankle sprains, which are rare, occur when the deltoid ligament is injured during excessive eversion. The foot rolls outward, and the patient lands on the inside of the foot. Because the lateral malleolus extends further distally than the medial malleolus, the ankle has a smaller range of eversion than inversion, which accounts for the more common occurrence of lateral ankle sprains. Deltoid ligament tears are typically associated with ankle fractures (Lynch, 2002, p. 410).

High ankle sprains, or syndesmotic sprains, are far less common than either lateral or medial sprains, accounting for only 1 to 11% of ankle sprains (Small, 2009, p. 317). High ankle sprains manifest themselves in the separation of the tibia and the fibula, or the widening of the ankle mortise. External rotation and hyperdorsiflexion are the most common causes of high ankle sprains. External rotation causes injury to the tibiofibular ligaments, allowing the tibia and fibula to separate; hyperdorsiflexion causes the wide anterior portion of the talus to push the malleoli apart. High sprains are common in collision sports including football, hockey, and soccer, as well

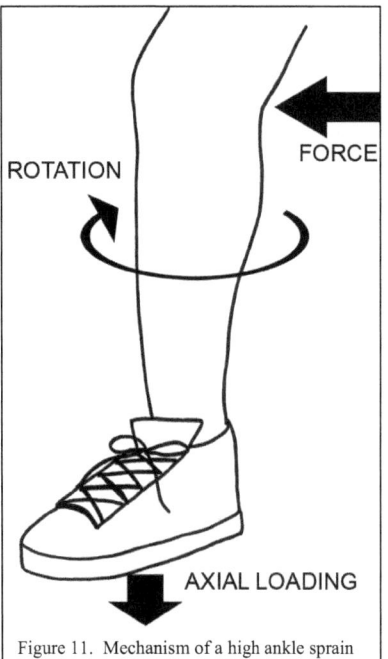

Figure 11. Mechanism of a high ankle sprain

as in skiing. A blow to the lateral leg while the foot is planted, or a ski that sticks in the snow while turning, can result in external rotation (Figure 10). Hyperdorsiflexion can occur when a hockey player's skate is forced into the boards, or when a runner comes to a sudden stop with the foot planted and falls forward (Figure 11) (Molinari et al., 2009, p. 132; Norkus et al., 2001, p. 71-72).

# Ankle Fractures

Ankle fractures are classified as unimalleolar, bimalleolar, or trimalleolar. Unimalleolar fractures involve injuries to either the medial or lateral malleolus. Bimalleolar fractures involve injuries to both the medial and lateral malleoli. A bimalleolar equivalent fracture occurs when the lateral malleolus is fractured and the deltoid ligament is completely ruptured. Trimalleolar fractures involve a combination of both medial and lateral malleollar fractures and either a posterior malleolar or a posterior tibial fracture.

Mechanisms of unimalleolar, bimalleolar, and trimalleolar fractures include falling forward on top of the foot, falling onto the outside of the foot while the foot is planted (pronation-abduction), and excessive rotational force while the foot is planted. Pilon fractures (fractures of the articular surface of the distal end of the tibia) are caused by axial loading mechanisms (ie., falling from a considerable height and landing on one's feet).

Ankle fractures are classified using two systems: the Danis-Weber classification and the Lauge-Hansen classification. The Danis-Weber classification system is based on the level of the fibula fracture. Type A fractures occur distal to the joint line or ankle mortise (i.e., fractures below the ankle joint); type B fractures occur at the level of the ankle joint; and type C fractures occur proximal to (or above) the level of the ankle joint.

The Lauge-Hansen system is based on the mechanism of injury and pathology: the first part identifies the position of the ankle at the time of injury, and the second part identifies the type of force applied to the ankle. There are four Lauge-Hansen classification groups: supination-adduction, supination-external rotation, pronation-abduction, and pronation-external rotation (Small, 2009 p. 317-319).

## Ankle Arthritis

Arthritis refers to inflammatory joint disease. The presenting symptoms of arthritic joints include pain, swelling, stiffness, redness of the skin about the joint, effusion, deformity, and ankylosis (Rogers, 2011, p. 218). Although the ankle joint is the most commonly injured joint of the human body, symptomatic arthritis of the ankle is nine times less likely than that at the knee and hip, and the ankle joint is rarely affected by osteoarthritis. The most common causes of the degenerative changes of ankle arthritis are traumatic injuries (fractures of the malleoli, tibial plafond, and talus and osteochondral damage of the talar dome) and abnormal ankle mechanics (ankle instability due to chronic lateral ligament laxity) (Thomas & Daniels, 2003, p. 923).

## Tendonitis of the Ankle

Tendonitis (tendinitis) is the inflammation of a tendon, and can affect the ankle joint when any of the tendons associated with the ankle joint, including the Achilles tendon, become inflamed. Achilles tendonitis, as well as other Achilles tendon injuries, is common among runner and is caused by overuse, improper training, gait abnormalities, degenerative changes, and improper footwear (Dubin, 2005, p. 39).

## Bursitis of the Ankle Joint

Bursitis of the ankle joint is the inflammation of one, or more, of the bursae that surround the ankle joint. The retrocalcaneal bursa and the retroachilles bursa are common sites of inflammation. The most common cause of retrocalcaneal and retroachilles bursitis is ill-fitting footwear that irritates the area of the Achilles tendon insertion at the posterior calcaneus (Aldridge, 2004, p. 334).

## Mechanism of Injury vs. Injury Type

There are no absolute rules for positively associating a type of ankle injury with a specific mechanism of injury, and vice versa. For example, a supination-adduction mechanism of injury can result in a unimalleolar, bimalleolar, or trimalleolar fracture, depending on several variables such as the severity and the exact location of the application of the mechanism of injury. Furthermore, an ankle injury may be associated with a secondary injury. A working knowledge of the mechanisms of common ankle injuries, however, can lead one to postulate possible mechanisms for specific ankle injuries.

Imagine that a gymnast poorly dismounted from a balance beam, resulting in a trimalleolar fracture of the right ankle, which consisted of a spiral fracture of the distal fibula, an avulsion fracture of the posterior malleolus, and a transverse fracture of the medial malleolus. Trimalleolar fractures typically occur when there is excessive rotational force while the foot is planted. This gymnast's injury can be explained by a supination-external rotation (or supination-lateral rotation) mechanism of injury: when the gymnast landed from her dismount, she rolled her foot inward (supination) as the foot rotated to her right (external rotation). A rupture of the anterior inferior tibiofibular ligament will also be associated with this stage IV supination-external rotation injury (Arimoto & Forrester, 1980, p. 1060).

## Integrative Medicine Therapies

Integrative medicine is a "whole-person" approach to treating a patient, designed to treat the person and not just the disease (in this case, the ankle injury), by treating the mind, body, and spirit. As every patient is unique, a thorough understanding of the individual patient is essential and can be achieved through diet journals, patient interviews, and lab testing. Most integrative medicine programs combine conventional Western medicine with alternative or complementary treatments and therapy, including herbal medicine, acupuncture, prolotherapy, massage, biofeedback, yoga, and stress reduction techniques. Duke Integrative Medicine's website states that through a partnership between the patient, physician, and a team of clinical experts, integrative medicine also anticipates possible health issues or risks and promotes prevention to minimize them (Duke Integrative Medicine, 2011).

## Supplements and Herbs

According to the University of Maryland Medical Center, some nutrients and herbs help restore damaged tissue, reduce swelling, and provide pain relief: Vitamin C, beta-carotene, glucosamine, chondroitin, calcium, and magnesium promote the healing and rebuilding of tissues. Pain relief can be achieved with the use of willow bark, cat's claw, and devil's claw. Bromelain, licorice, white willow, Vitamin E, and essential fatty acids such as fish oil or primrose oil reduce inflammation. White willow, Aescin, and Tumeric reduce swelling. Vitamin C, beta-carotene, and Vitamin A help improve immune function (Ehrlich, March 7, 2010; March 29, 2010).

## Acupuncture

Acupuncture originated in China many centuries ago. Its use spread throughout Asia, and it was introduced to Europe in the 17th century. Acupuncture aims to heal through the stimulation of anatomical points on the body using a variety of techniques. Acupuncture involves penetrating the skin with thin, solid, metallic needles manipulated by the hands or by electrical stimulation. According to the World Health Organization (2003), other techniques associated with acupuncture include moxibustion (the burning on or over the skin of selected herbs), laser acupuncture, and acupressure. According to traditional Chinese medicine, acupuncture regulates the flow of qi, or vital energy, through the body, thus keeping the body in a balanced state (National Center for Complementary and Alternative Medicine, 2011). The World Health Organization compiled a review and analysis of controlled clinical trials of acupuncture therapy. The study concluded that acupuncture analgesia works better than a placebo for most kinds of pain, and is highly effective in treating chronically painful conditions. Acupuncture alleviates pain and reduces muscle spasm; for the treatment of sprains, acupuncture can also improve local circulation, thus speeding up the recovery time (World Health Organization, 2003).

## Homeopathy

The guiding principle of homeopathy is the principle of similars or "like cures like", in which a disease can be cured by a substance that produces similar symptoms in healthy people. Another important principle of homeopathic treatments is dilution: the lower the dose of the medication, the greater its effectiveness. Most homeopathic remedies are so dilute that no molecules of the healing substance remain; even so, the healing substance has left its imprint or "essence," and it is this essence which cures the disease. Homeopaths treat patients based on their genetic and personal history, body type, and current symptoms; therefore, remedies are individualized to each patient.

Homeopathic remedies are derived from natural substances that come from plants, minerals, or animals (National Center for Complementary and Alternative Medicine, 2010). Ehrlich (March 29, 2010), with the University of Maryland Medical Center, lists Arnica (topical or internal), Byronia, Ledum, Rhus toxicodendron, Ruta, and Traumeel as homeopathic remedies for sprains. Ehrlich (March 7, 2010) also lists Byronia, Phytolacca, Rhus toxicodendron, and Rhododendron, as well as injectable homeopathic medications such as Traumeel, as homeopathic remedies for tendonitis. In the treatment of acute injuries, Traumeel, an inflammation regulating drug, is often combined with Spascupreel (for muscle strains) and Lymphomyosot (for tissue swelling) (Barkauskas, 2007, p. 6).

A case study by Steven Rosenberg, D.P.M., (1998) related his homeopathic treatment of a 45-year-old woman with an ankle sprain. After a physical examination and evaluation of x-rays of the left ankle, Dr. Rosenberg diagnosed this patient with a 1st degree left ankle sprain. He treated the patient with multiple subcutaneous injections of Traumeel to all three ligament sites in the lateral aspect of the ankle. An Unna boot soft immobilization cast was applied to the left ankle to provide stability and compression. The patient was also given Traumeel (anti-inflammatory), Osteoheel (pain relief), and Lymphomyosot (edema relief) tablets to help decrease the pain, inflammation, and swelling, and Traumeel ointment for topical application. The following day, the patient had a less painful ankle that could support her weight. The swelling was decreasing, and she could flex her ankle without pain (p. 280). This case study shows how conventional techniques (application of the Unna boot) in concert with alternative medicine (homeopathic remedies) can effectively relieve the symptoms of ankle injuries.

## Prolotherapy

Prolotherapy involves a series of injections of irritants, osmotic shock agents, and/or chemotactic agents designed to stimulate low-grade inflammation in injured tissues, specifically ligaments, tendons, and cartilage, which promotes tissue repair and/or growth.

When tissues, such as ligaments, are injured, the common initial response is inflammation, which stimulates substances carried in blood that produce growth factors in the injured area to promote healing. Ligaments, tendons, and cartilage, however, have poor blood supply and take longer to heal than other tissues; as a result, incomplete healing of these structures is common. Traditional treatments for ligament and tendon injuries include anti-inflammatory medications (ibuprofen and Naprosyn), nonsteroidal anti-inflammatory drugs (NSAIDS), or corticosteroids to relieve pain and/or swelling to provide temporary relief. Gordin (2011) wrote that a study by Dr. Richard Wrenn demonstrated "suppressed fibroblastic reactions (connective tissue formation) to injury following intramuscular injections of cortisone" (p. 601). Proponents of prolotherapy argue that by suppressing inflammation and/or fibroblast proliferation and collagen formation, these traditional treatments actually suppress the body's natural healing process, and the injured tissues do not fully heal. As a result, many patients suffer from chronic ankle sprains, laxity, or instability due to incomplete healing.

According to Alderman (2007), the use of prolotherapy techniques dates back to Ancient Greece and Hippocrates, who used red-hot needle cautery to treat dislocated shoulders, but the term "Prolotherapy" was by George S. Hackett, M.D., in 1956 as "the rehabilitation of an incompetent structure [ligament or tendon] by the generation of new cellular tissue" (p. 10-11). Gordin (2011) states that common proliferant solutions used in prolotherapy treatments include dextrose, glycerin, minerals, sodium morrhuate, autologous growth factors, and other pro-inflammatory compounds. These injected substances irritate the injured ligaments, tendons, or cartilage, which stimulates the formation of collagen (the major component of connective tissue, i.e. ligaments and tendons) via the production of fibroblasts (cells that synthesizes collagen), resulting in tissue growth and repair of injured structures (p. 605-606).

Numerous studies have demonstrated the development and growth of new ligamentous tissue in joints throughout the body using prolotherapy treatment. A case study by Clive Sinoff, M.D., (2010), documented prolotherapy treatment of a 58-year old man with a 20-year-old ankle injury. This gentleman injured both ankles due to a fall off a roof. He had physical therapy for 7 years, underwent arthrodesis of both ankles and the right foot, and utilized therapeutic ultrasound, a TENS unit, and hot foot soaks to only transient, mild relief. Prolotherapy treatments were started in May 2006 and ended in August 2008 (a total of seven sessions over two years). By

July 2009, the patient reported minimal pain and no longer needed analgesics (p. 487-488).

## Summary

Although the anatomy of the ankle is rather complex, a careful examination of the individual components of the ankle makes the entire structure easier to understand. The same process applies to ankle injuries: once the specific functions of the components of the ankle are examined, the mechanisms of injury to those same components become clear. Through this examination of the ankle and its mechanisms of injury, it is easy to see why people suffer ankle injuries when applied forces cause the components of the ankle to exceed their physical limits. Additionally, through Integrative Medicine, treatment of ankle injuries (and symptoms) can be performed in concert with treatment of the whole patient using alternative and complementary techniques.

# References

Alderman, D. (2007, January/February). Prolotherapy for musculoskeletal pain. *Practical Pain Management, 7* (1), 10-15. Retrieved from http://www.prolotherapy.com/ppm2007.pdf

Aldridge, M.D. (2004). Diagnosing heel pain in adults. *American Family Physician, 70* (2), 332-338.

Ardizzone, R., & Valmassy, R.L. (2005). How to diagnose lateral ankle injuries. *Podiatry Today,* 18 (10), 65-74.

Arimoto, H. K., & Forrester, D. M. (1980). Classification of ankle fractures: an algorithm. *American Journal of Roentgenology,* 135, 1057-1063.

Arnheim, D. D., & Prentice, W. E. (1993). *Principles of Athletic Training (8th ed.).* St. Louis: Mosby Year Book.

Barkauskas, D. (2007). Treating sports injuries—a functional approach. *Journal of Biomedical Therapy,* 1(1), 4-8.

Brody, J. E. (2007, August 7). Injections to kick-start tissue repair. *The New York Times.* Retrieved from http://www.nytimes.com/2007/08/07/health/07brod.html?_r=2&pagewanted

Dox, I.G., Melloni, B.J., & Eisner, G.M. (1993). *The Harper Collins Illustrated Medical Dictionary.* New York: HarperCollins.

Dubin, D.C. (2005) Athletes strain to avoid Achilles tendon problems. *Biomechanics,* http://www.biomech.com/full_article/?ArticleID=645&month=08&year=2005

Duke Integrative Medicine. (2011). What is Integrative Medicine? Retrieved from http://www.dukeintegrativemedicine.org/about-us/what-is-integrative-medicine

Ehrlich, S. D. (2010, March 7). Tendinitis. Retrieved from http://www.umm.edu/altmed/articles/tendinitis-000163.htm

Ehrlich, S. D. (2010, March 29). Sprains and Strains. Retrieved from http://www.umm.edu/altmed/articles/sprains-and-000157.htm

Fong, D.T.P., Chan, Y., Mok, K., Yung, P.S.H., & Chan, K. (2009). Understanding acute ankle ligamentous sprain injury in sports. *Sports Medicine, Arthroscopy, Rehabilitation, Therapy & Technology*, 1 (14). doi: 10.1186/1758-2555-1-14.

Gordin, K. (2011, May). Case for prolotherapy. *Journal of Prolotherapy*, 3 (2), 601-609. Retrieved from http://www.journalofprolotherapy.com/pdfs/issue_10/JOP_vol_3_issue_2_may_2011.pdf#page=11

Kapit, W., & Elson, L. M. (1993). *The Anatomy Coloring Book (2nd ed.).* New York: Harper Collins College Publishers.

Lynch, S.A. (2002). Assessment of the injured ankle in the athlete. *Journal of Athletic Training*, 37 (4), 406-412.

Martini, F.H., & Bartholomew, E.F. (2000). *Essentials of Anatomy & Physiology, 2nd Ed.* Upper Saddle River, New Jersey: Prentice-Hall.

McMinn, R.M.H., Hutchings, R.T., Pegington, J., & Abrahams, P. (1996). *Color Atlas of Human Anatomy, 3rd Ed.* London: Mosby-Wolfe.

Molinari, A., Stolley, M., Amendola, A. (2009). High ankle sprains (syndesmotic) in athletes: diagnostic challenges and review of the literature. *The Iowa Orthopaedic Journal*, 29, 130-138.

Morrison, K.E., & Kaminski, T.W. (2007). Foot characteristics in association with inversion ankle injury. *Journal of Athletic Training,* 42 (1), 135-142.

National Center for Complementary and Alternative Medicine. (2010, August). Homeopathy: An Introduction. Retrieved from http://nccam.nih.gov/health/homeopathy

National Center for Complementary and Alternative Medicine. (2011, August). Acupuncture: An Introduction. Retrieved from http://nccam.nih.gov/health/acupuncture/introduction.htm

Norkus, S.A., & Floyd, R.T. (2001). The anatomy and mechanisms of syndesmotic ankle Sprains. *Journal of Athletic Training*, 36 (1), 68-73.

Patton, K.T., & Thibodeau, G.A. (2000). *Mosby's Handbook of Anatomy & Physiology.* St. Louis: Mosby.

Rogers, K. (2011). *Bone and Muscle: Structure, Force, and Motion.* New York: Britannica Educational Publishing.

Rosenberg, S. (1998). Homeopathic treatment of ankle sprain: a case study. *Biomedical Therapy*, 16(4), 280.

Sinoff, C. (2010, November). Prolotherapy for 20 year old ankle injury. Journal of Prolotherapy, 2 (4), 497-488. Retrieved from http://www.journalofprolotherapy.com/pdfs/issue_08/issue_08_05_old_ankle_i njury.pdf

Small, K. (2009). Ankle sprains and fractures in adults. *Orthopaedic Nursing*, 28 (6), 314-320.

Thomas, R.H., & Daniels, T.R. (2003). Ankle arthritis. *Journal of Bone and Joint Surgery*, 85 (5), 923-926.

Wolfe, M.W., Uhl, T.L., Mattacola, C.G., & McCluskey, L.C. (2001). Management of ankle sprains. *American Family Physician*, 63, 93-104.

World Health Organization. (2003). Acupuncture: Review and Analysis of Reports on Controlled Clinical Trials. Retrieved from http://apps.who.int/ medicinedocs/en/d/Js4926e/

# Chapter 3

## Comparison of Forward vs. Sideways Slip-and-Fall Mechanisms

The Slip-and-Fall is a common cause of personal injury. Understanding the mechanics of slip-and-fall accidents can help to illuminate the relationship (or lack thereof) between a slip-and-fall incident and any subsequent injury complaints. In particular, we are interested in exploring the relationship (or lack thereof) between a slip-and-fall accident and an ankle injury. The majority of slip-and-fall accidents occur as one is walking in a straight line; we identify this mechanism as a "forward slip-and-fall." A minority of slip-and-fall accidents, however, can be classified as "sideways slip-and-fall" accidents. This mechanism can be commonly seen when the slip occurs during a turning or pivoting maneuver.

In a forward slip-and-fall accident, the slip initiates at heel strike. The leading foot slips forward and the upper body thrusts backward, typically resulting in a fall backwards and/or to the side. There are variations to the forward slip-and-fall accident: if the trailing leg and foot cannot swing forward during the fall, the trailing knee may make first contact with the floor, or the person may land on top of a bent trailing leg. In a forward slip-and-fall accident, we would not expect any ankle injury to the leading foot, as it is highly unlikely that the leading foot would experience any twisting or rotating mechanism consistent with ankle injury. Figures 1 and 2 illustrate the initiating moment of a forward Slip-and-Fall (front and side views). Figure 3 depicts the bones and ligaments of the ankle at the initiating moment of a forward Slip-and-Fall.

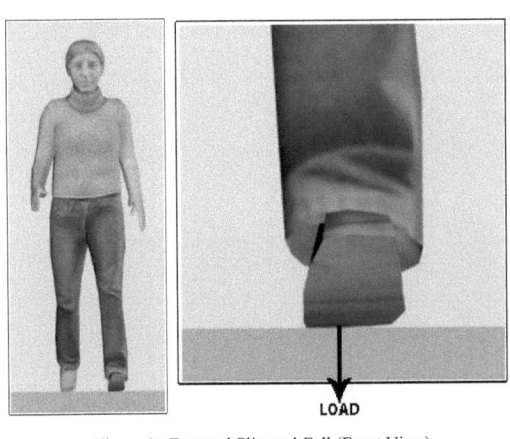

LOAD

Figure 1. Forward Slip-and-Fall (Front View)

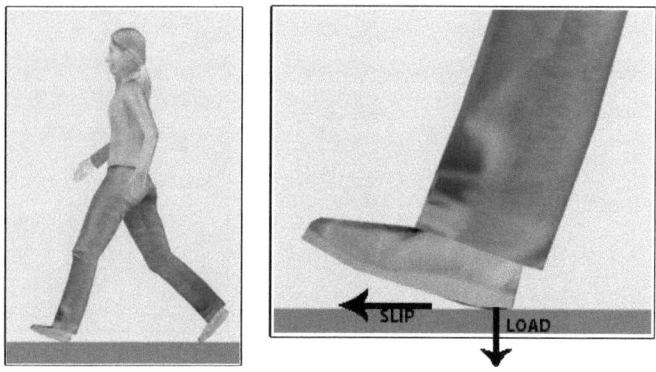

Figure 2.  Forward Slip-and-Fall (Side View)

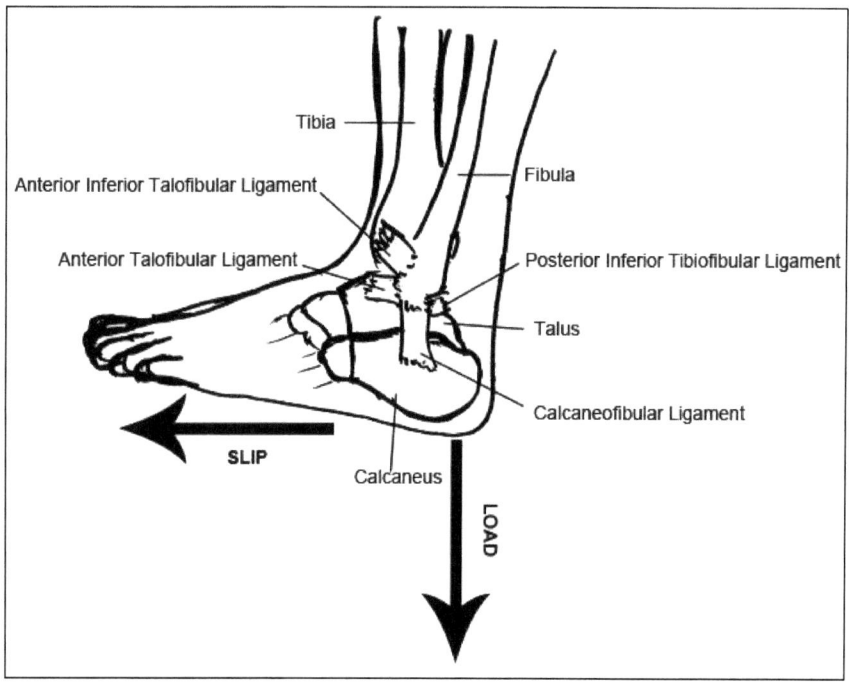

Figure 3.  Bones and ligaments of the ankle during a forward slip-and-fall accident

In a sideways Slip-and-Fall, the slip commonly initiates during a turning or pivoting maneuver. As the foot pivots, the foot is loading both longitudinally and laterally. If a slip initiates during a pivot or turn, the foot will slip sideways, with a high

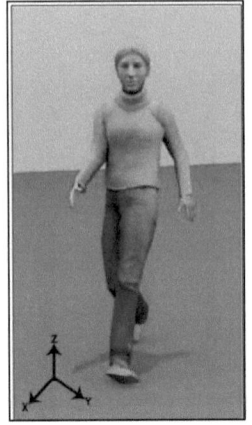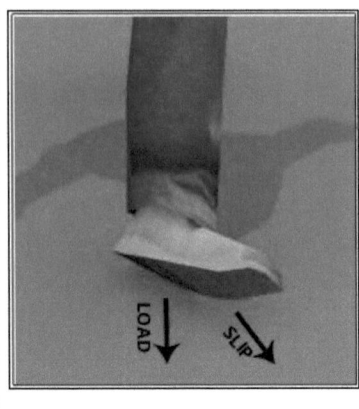

Figure 4. Sideways Slip-and-Fall (Front View)

probability of inversion or eversion of the leading foot. We expect that the body will fall sideways. Due to the presence of a twisting mechanism (inversion/eversion), an ankle injury would be expected with a sideways slip-and-fall accident. Figures 4 and 5 illustrate

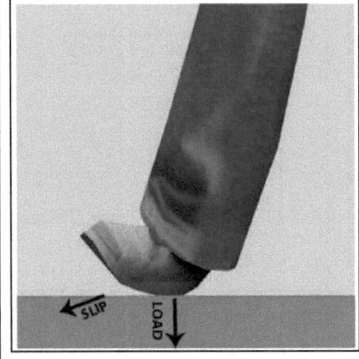

Figure 5. Sideways Slip-and-Fall (Side View)

the initiating moment of a sideways Slip-and-Fall resulting from a left foot slip during a left turn maneuver (front and side views). Figure 6 illustrates the bones and ligaments of the ankle during a sideways Slip-and-Fall (including the anticipated ankle inversion).

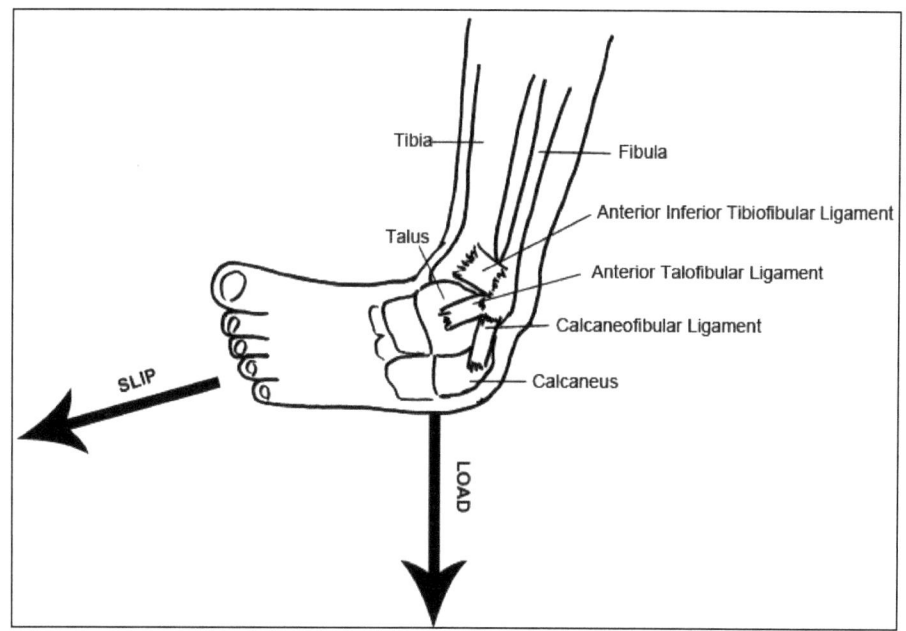

Figure 6. Bones and ligaments of the ankle during a sideways slip-and-fall accident

As discussed above, a thorough understanding of the mechanics of a slip-and-fall accident can help to identify related injuries. Through the examples above, we investigated the possibility of an ankle injury as a result of a slip-and-fall accident via an examination of two possible slip-and-fall mechanisms (forward and sideways). A similar study may be conducted to investigate the probability of injury to other parts of the body (i.e., hips, knees, wrists, spine, etc.) as a result of slip-and-fall accidents.

# Chapter 4

## Shoulder Injury Mechanisms and Integrative Medicine Therapies

(Originally published in the American Association of Integrative Medicine Blog,
retrieved from http://www.aaimedicine.com/blog/2014/08/shoulder-injury-
mechanisms-and-integrative-medicine-therapies/)

## Introduction

The purpose of this paper is to discuss the relationship between several different
mechanisms of shoulder injury and the types of injury that commonly occur as a
result. While there are no absolute rules for linking a single mechanism of injury
with the specific type of injury it will produce, or vice versa, this paper attempts to
illuminate the relationship between mechanism and injury, as well as the relationship
between associated injuries. Furthermore, we will discuss common Integrative
Health treatments for patients suffering from shoulder injuries. We begin by
providing a preliminary examination of the anatomy of the shoulder, as well as the
range of motion of the shoulder. Next we discuss the common mechanisms of injury
to the shoulder, followed by a discussion of the types of shoulder injuries that arise
from said mechanisms. We continue by illustrating the relationships between
associated injuries. And finally, we will discuss common integrative health therapies
for shoulder injuries.

## Shoulder Anatomy

The shoulder is the most mobile joint in the human body; it allows rotation of the
upper extremity up to 180 degrees in three planes. A downside to the versatility of
movement in the shoulder is an increased risk of injury, as it is the most frequently
dislocated joint (Martini, 2000, p. 156; Quillen, 2004, p. 1947). The shoulder girdle
is comprised of the clavicle, the scapula, and the proximal end of the humerus—the

bone of the upper arm (Figures 1 and 2). The articular surface of the humeral head is hemispherical. The clavicle, or collar bone, is a thin, curved bone that extends transversely from the sternum to the acromial end of the scapula. It provides the only bony connection between the axial skeleton and the upper extremity. Clavicle fractures are classified based on the segment where the fracture occurs: Zone 1

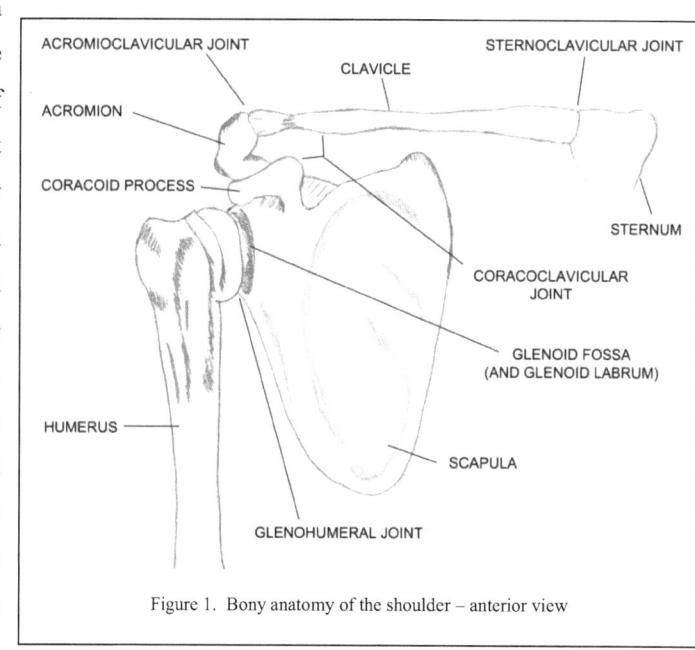

Figure 1. Bony anatomy of the shoulder – anterior view

fractures, which make up 80% of all clavicular fractures, occur in the middle $^1/_3$ of the clavicle; Zone 2 fractures, which consist of 15% of all clavicular fractures, occur in the distal $^1/_3$ of the clavicle; and Zone 3 fractures, the least common at 5% of all clavicular fractures, occur in the proximal $^1/_3$ of the clavicle (Hannon, 2006, p. 240-241; Quillen, 2004, p. 1947-1948).

The scapula, or shoulder blade, is a thin, triangular-shaped bone lying on the posterior and lateral aspect of the thorax, which is stabilized by the coracoclavicular ligaments and muscular attachments; it has no direct bony attachment to the axial skeleton. The scapula has several shallow depressions, or fossae (the infraspinous fossa, supraspinous fossa, and the subscapular fossa). The scapula's spine, which crosses its posterior surface obliquely, ends in the acromion, the process that articulates with the clavicle. The lateral apex of the scapula is broadened and

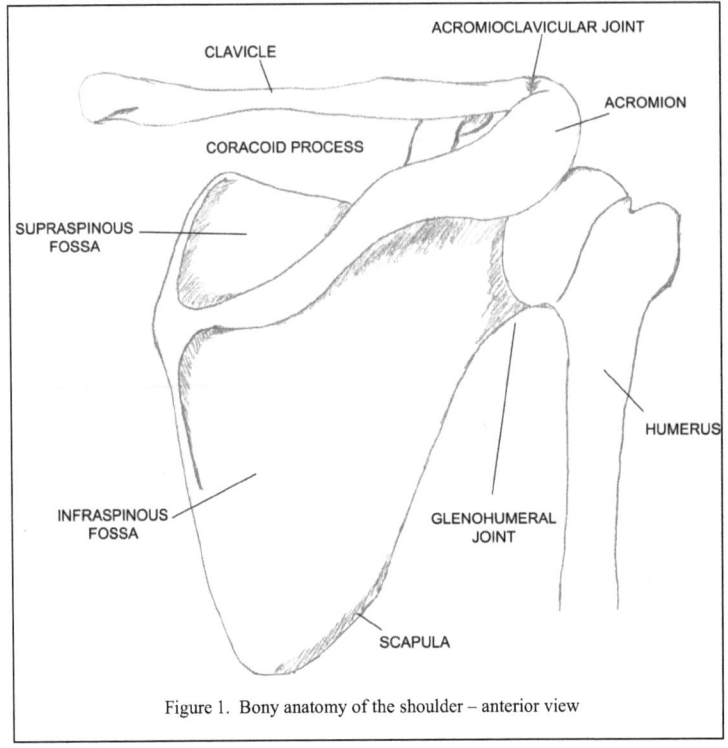

Figure 1. Bony anatomy of the shoulder – anterior view

presents a shallow cavity, the glenoid fossa, which articulates with the humeral head. The coracoid process is a beaklike projection that hangs over the glenoid cavity. It can be felt in the groove between the deltoid and the pectoralis muscles and 1 inch below the clavicle (Hannon, 2006, p. 237; Martini, 2000, p. 142-144; Patton, 2000, p. 138; Rogers, 2011, p. 79-80).

## Joints of the Shoulder

There are four joints and one articulation (the scapulothoracic articulation), in the shoulder. The joints of the shoulder girdle include the sternoclavicular joint, the acromioclavicular joint, the coracoclavicular joint, and the glenohumeral joint (Figures 1, 2, and 3). The scapulothoracic "joint," or articulation, is formed only by muscular connections and involves no ligamentous attachments. The coracoclavicular joint is a syndesmotic joint (an articulation between two bones joined by a ligament) found between the coracoid process of the scapula and the clavicle.

The sternoclavicular joint is formed by the sternum and the clavicle; it is the only point at which the shoulder girdle is attached to the axial skeleton. It is a saddle-shaped synovial joint with an intra-articular disc. This joint is inherently unstable; the costoclavicular ligament, which binds the inferior surface of the clavicle to the superior surface of the first costal cartilage and first rib, is the major stabilizer of the sternoclavicular joint. Movements of the sternoclavicular joint include elevation, depression, protrusion, retraction, and rotation (Christensen, 2002; Hannon, 2006, p.237).

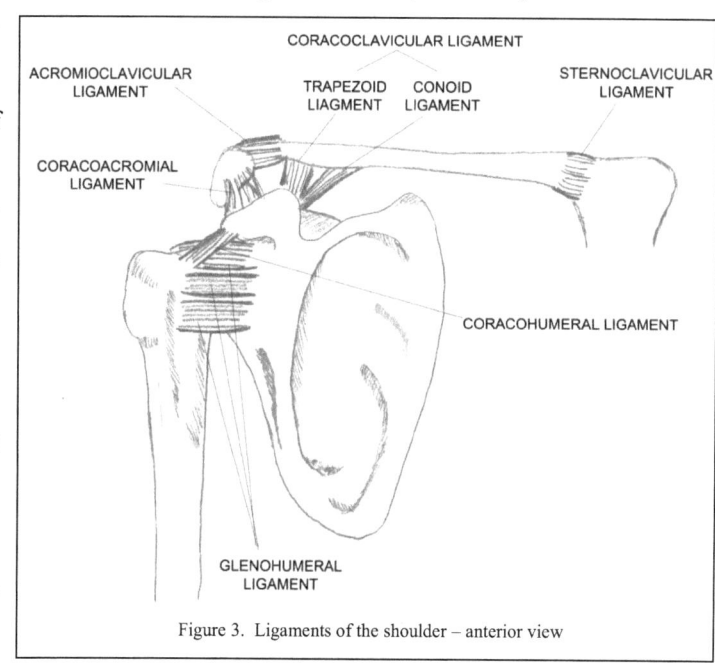

Figure 3. Ligaments of the shoulder – anterior view

The acromioclavicular (AC) joint, a plane synovial joint, is formed by the medial side of the acromion of the scapula and the lateral end of the clavicle. The superior and inferior acromioclavicular ligaments prevent the clavicle from overriding the acromion. The coracoclavicular ligaments are secondary stabilizers of the acromioclavicular joint, providing a syndesmotic junction between the clavicle and the scapula (Fialka, 2004, p. 21). The acromioclavicular joint allows a longitudinal rotation of about 40 degrees (Fialka, 2004, p. 21). The Rockwood classification for acromioclavicular joint sprains, or separations, includes six categories: Type I sprains consist of tenderness over the AC joint, but no visible deformity of the distal clavicle. Type II sprains consist of slight widening of the AC joint, a more prominent distal clavicle, and possible pain at the distal end of the clavicle due to a sprained coracoclavicular ligament. Type III sprains have an obvious visible prominence of the distal clavicle which is due both to the separation of the acromioclavicular joint and an increase of the coracoclavicular distance. Types IV through VI each have grossly abnormal radiographs. Type IV sprains have a lateral clavicle displaced posteriorly and perforating the trapezius muscle. Type VI is defined as a subcoracoidal displacement of the clavicle. Type I and II sprains, and some Type III sprains, can be treated nonoperatively. AC Sprain Types IV through VI must be treated operatively (Fialka, 2004, p. 21; Quillen, 2004, p. 1951-1952).

The glenohumeral joint is formed by the head of the humerus and the glenoid fossa of the scapula. It is a multiaxial ball-and-socket synovial joint which is supported by muscles (including the rotator cuff muscles), tendons, capsular tissue, and the glenoid labrum for stability. The connecting ends of the bones are surrounded by a joint capsule lined with a synovial membrane and containing synovial fluid. The large size of the humeral head is able to move unrestricted on the small glenoid cavity, and the laxity of the capsule allows the humerus to move easily. This allows a wide range of movements for the humerus, which include adduction, abduction, extension, flexion, internal rotation, external rotation, horizontal abduction, horizontal adduction, and circumduction. The two main ligaments of this joint are the glenohumeral ligaments (superior, middle, and inferior) and the coracohumeral ligament (Hannon, 2006, p. 238; Kapit, 2002, p. 31-32; Woodward, 2000, p. 3079).

## The Glenoid Labrum

The glenoid labrum, part of the glenohumeral joint, is a cuff of fibrocartilaginous tissue that surrounds the glenoid cavity. It deepens the glenoid fossa of the scapula and increases the surface area of the articulation with the humeral head, thus increasing joint stability. It also allows the attachment of the tendon of the long head of the biceps brachii muscle and the glenohumeral ligaments. The superior and anterosuperior portions of the labrum are loosely attached to the glenoid fossa. The biceps tendon inserts into the superior portion of the labrum. A common tear of the glenoid labrum is the SLAP (superior labral anteroposterior) tear. SLAP tears are usually centered at the attachment of the biceps tendon and extend to either the anterior or posterior portion of the labrum. Anterior glenohumeral dislocations are commonly associated with both SLAP tears and either Bankart lesions (injuries to the anterior glenoid labrum) or Hill-Sachs lesions (a bony indentation at the back of the humeral head) (Cutts, 2009, p. 3; Hannon, 2006, p. 238; Mohana-Borges, 2003, p. 1454).

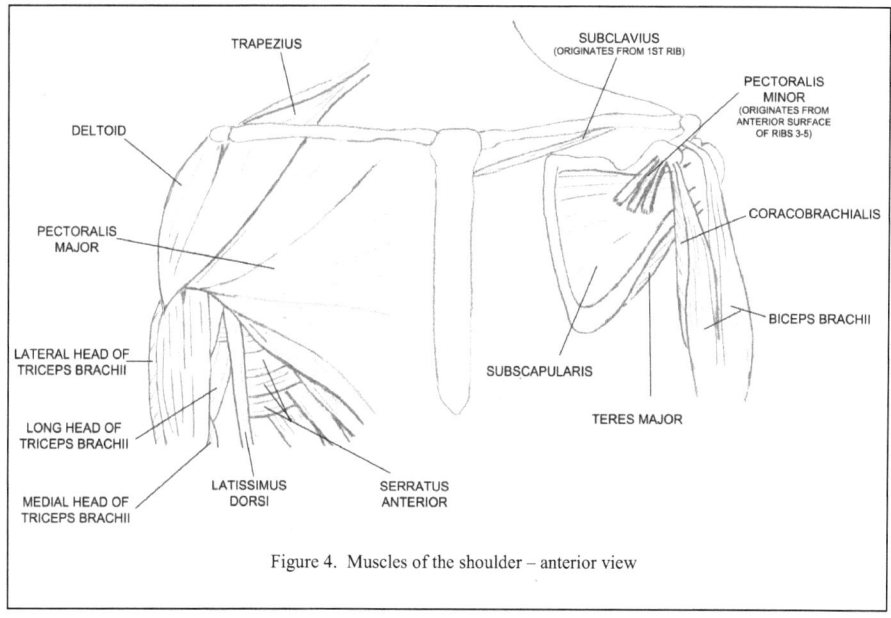

Figure 4. Muscles of the shoulder – anterior view

## The Rotator Cuff

The rotator cuff is the dynamic stabilizer of the glenohumeral joint; it maintains proper position of the humeral head within the glenoid fossa during shoulder movements. The rotator cuff (also referred to as the SITS muscles) is comprised of four muscles: the subscapularis, supraspinatus, infraspinatus, and teres minor muscles (Figures 4 and 5) (Kapit, 2002, p. 55). Without an intact rotator cuff, the unopposed deltoid muscle would cause cephalad migration of the humeral head, resulting in subacromial impingement of the rotator cuff (Fongemie, 1998, p. 668).

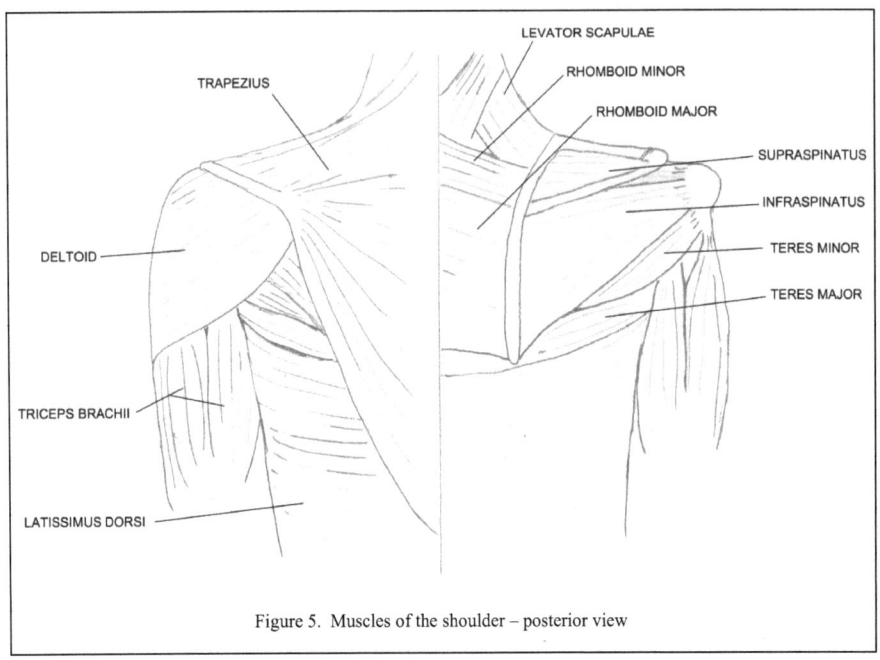

Figure 5. Muscles of the shoulder – posterior view

## The Impingement Interval

The space between the undersurface of the acromion and the superior aspect of the humeral head is called the subacromial space, or impingement interval. The height of the space between the acromion and the humeral head ranges from 1.0 to 1.5 centimeters; however, between the two structures lie the rotator cuff tendons, the long head of the biceps tendon, the subacromial bursa, and the coracoacromial ligament, which decreases that distance (Bigliani, 1997, p. 1855). The impingement interval is maximally narrowed with abduction. Further narrowing of this space, whether by chronic overuse and wear-and-tear or by traumatic causation, can cause an increase in pressure within the confined space, resulting in impingement syndrome. There are three types of anatomical configuration of the acromion process of the scapula that affect the shape and size of the impingement interval, and possibly contribute to impingement syndrome: type I acromions have the "normal" configuration, which is flat; type II acromions are curved and dip downward; and type III acromions have a hooked shape and dip downward, obstructing the outlet for the supraspinatus tendon (Fongemie, 1998, p. 668-669).

## Brachial Plexus

The brachial plexus is comprised of nerve roots which originate from the cervical and thoracic spine (C5 to T1) and split off into trunks, cords, and branches, which travel through the shoulder and innervate the muscles of the shoulder, arm, forearm, hand, and fingers (Kapit, 2002, p. 88). The neurovascular bundle runs through three narrow passageways from the base of the neck toward the axilla and the proximal arm. One of these passageways, the interscalene triangle, is formed by the anterior scalene muscle anteriorly, the middle scalene muscle posteriorly, and the medial surface of the first rib inferiorly, and contains the trunks of the brachial plexus and the subclavian artery. Thoracic outlet syndrome is the result of compression of the brachial plexus or subclavian vessels in the thoracic outlet. The majority of TOS cases involve neural and/or vascular compression within the interscalene triangle. TOS can be produced by bone malformation, fibromuscular anomalies, or scarring following trauma to the neck and/or shoulders (Hannon, 2006, p. 239; Huang, 2004, p. 898).

## Bursae

A bursa is a fluid-filled sac between tendons, muscles, or skin and bony prominences at points of friction or stress. Inflammation of one of these bursae, located over a joint or between tendons and muscles or bones, is called bursitis. Inflammation of the subacromial bursa, which lies just below the acromion, is commonly associated with impingement syndrome (Martini, 2000, p. 156; Rogers, 2011, p. 139, 220).

**Table #1. Muscles and Movements of the Shoulder**

| Action | Muscles Involved |
|---|---|
| Scapular Adduction (Retraction) | Rhomboid major, rhomboid minor, trapezius |
| Scapular Abduction (Protraction) | Serratus anterior, pectoralis minor, subclavius |
| Scapular Elevation | Levator scapulae, trapezius |
| Scapular Depression | Pectoralis minor, trapezius, subclavius |
| Arm Adduction | Deltoid, pectoralis major, latissimus dorsi, teres major, coracobrachialis, triceps brachii |
| Arm Abduction | Abduction to 120 degrees: supraspinatus, deltoid Upward rotation of scapula: trapezius, serratus anterior |
| Flexion | Pectoralis major, coracobrachialis, biceps brachii, deltoid |
| Extension | Latissimus dorsi, teres major, triceps, deltoid |
| Medial Rotation | Subscapularis, latissimus dorsi, teres major, pectoralis major, deltoid |
| Lateral Rotation | Infraspinatus, teres minor, deltoid |
| Circumduction | See muscles of flexion, abduction, extension, and adduction |

## Muscles of the Shoulder

There are 18 muscles which work together to perform the movements of the shoulder. These muscles are the trapezius, latissimus dorsi, teres major, teres minor, levator scapula, rhomboid major, rhomboid minor, deltoid, coracobrachialis, supraspinatus, infraspinatus, subscapularis, pectoralis major, pectoralis minor, subclavius, serratus anterior, biceps brachii, and triceps brachii (Figures 4 and 5). The actions of these muscles are summarized both in Table #1 and below.

## Kinematics (Movements) of the Shoulder

The shoulder, which is largely a multiaxial ball-and-socket joint, has a very large range of motion, which is achieved by the following movements (Kapit, 2002, p. 54-56; Martini, 2000, p. 196-197; McMinn, 1996, p. 120; Patton, 2000, p. 203, 206-207; Zuidema, 1997, p. 14):

During scapular adduction (scapular retraction), the scapula moves posteriorly and medially along the back, which moves the arm and shoulder joint posteriorly. This action is often referred to as "squeezing the shoulder blades together." The rhomboid major, rhomboid minor, and trapezius muscles play a role in scapular adduction.

During scapular abduction (scapular protraction), the scapula moves anteriorly and laterally along the back, moving the arm and shoulder joint anteriorly. This motion is achieved by "rounding one's back." The serratus anterior, pectoralis minor and subclavius muscles perform scapular abduction.

Scapular elevation raises the scapula, i.e., shrugging one's shoulders. The levator scapulae and upper fibers of the trapezius perform scapular elevation.

Scapular depression lowers the scapula, i.e., slumped shoulders. The muscles involved in scapular depression include the pectoralis minor, lower fibers of the trapezius, and subclavius.

In arm abduction, the arms are raised in the plane of the torso laterally. During true abduction, the humerus goes from parallel to the spine to perpendicular to the spine. The supraspinatus and deltoid muscles perform this motion. During upward rotation of the scapula, the humerus is raised to above the shoulders and straight upwards. The trapezius and serratus anterior muscles perform this action.

Arm adduction is performed by the deltoid, pectoralis major, latissimus dorsi, teres major, coracobrachialis, and triceps brachii muscles.

During arm flexion, the humerus is rotated out of the plane of the torso, so that it points anteriorly. The pectoralis major, coracobrachialis, biceps brachii, and the anterior fibers of deltoid play a role in arm flexion.

During arm extension, the humerus is rotated out of the plane of the torso, so that it points posteriorly. The latissimus dorsi and teres major, long head of triceps, and the posterior fibers of the deltoid play a role in arm extension.

The subscapularis, latissimus dorsi, teres major, pectoralis major, and the anterior fibers of the deltoid perform medial rotation of the arm.

The infraspinatus, teres minor, and the posterior fibers of the deltoid perform lateral rotation of the arm.

Arm circumduction is a complex movement of the shoulder in a circular motion, thus moving the arm in a loop. It is characterized as flexion, abduction, extension, and adduction done in sequence; thus, the muscles which perform flexion, abduction, extension, and adduction also perform the action of circumduction.

### Table #2.  Common Mechanisms of Shoulder Injuries

| Mechanism of injury | Possible injuries |
|---|---|
| Fall onto outstretched arm (external rotation and abduction) | Rotator cuff tear, clavicle fracture, proximal humerus fracture, anterior dislocation, subluxation, AC joint sprain/separation, SLAP tear |
| Break fall with hand (forces induced up the arm) | Clavicle fracture, posterior dislocation, proximal humerus fracture |
| Avulsion (pull or yank on arm) | Rotator cuff tear, SLAP tear |
| Sudden overhead movement of the arm | Rotator cuff tear, impingement, SLAP tear |
| Direct blow to shoulder | Clavicle fracture, proximal humerus fracture, anterior dislocation, posterior dislocation, AC joint sprain/separation, brachial plexus injury, sternoclavicular joint separation, bursitis |
| Fall onto shoulder | AC joint separation, brachial plexus injury, proximal humerus fracture, clavicle fracture |
| Traction injury (tilts head/neck to side and depresses shoulder) | Subluxation, brachial plexus injury |
| Swinging the arm strenuously into an awkward position | Subluxation |
| Overuse or repetitive motions | Rotator cuff tear, rotator cuff tendinopathy, glenoid labrum tear, biceps brachii tendon tear, bursitis, impingement |
| Degenerative or congenital conditions | Osteoarthritis, impingement, degenerative tendinopathy |

## Common Mechanisms of Shoulder Injury

The following are common mechanisms of shoulder injury and the possible injuries which may arise as a result (also summarized in Table #2):

A fall onto outstretched arm produces violent external rotation and abduction of the shoulder (Figure 6). Injuries which may arise include rotator cuff tears, clavicle fractures, proximal humerus fractures, anterior dislocations, subluxation, AC joint sprain/separations, and SLAP tears (Christensen, 2002;

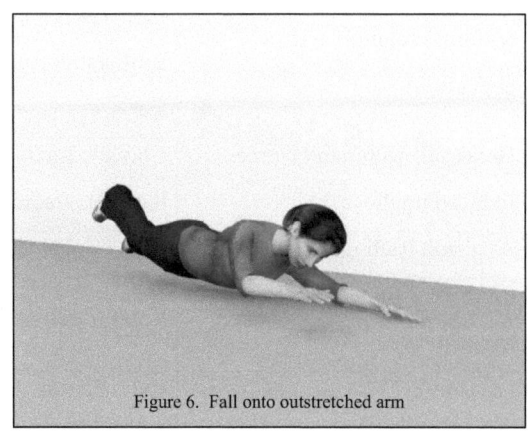

Figure 6. Fall onto outstretched arm

Hannon, 2006, p. 239; Jones, 2010, p. 66-68, 74; Mohana-Borges, 2003, p. 1450, 1454; Quillen, 2004, p. 1947).

Avulsion, which is a sudden pull or yank of the arm, can result in a rotator cuff tear or a SLAP tear. Sudden overhead movements of the arm (commonly seen in sports such as volleyball, baseball, swimming, and tennis) can cause rotator cuff tears, impingement (technically due to muscle weakness in the rotator cuff muscles caused by the tension overload seen when the arm is in an overhead position), and SLAP tears (Bigliani, 1997, p. 1855; (Mohana-Borges, 2003, p. 1450, 1454).

A direct blow to the shoulder can result in any of the following injuries: clavicle fracture, proximal humerus fracture, anterior dislocation, posterior dislocation (caused by a posteriorly directed force on the humeral head), AC joint sprain/separation (caused by a direct blow to the acromion with the humerus in the adducted position), brachial plexus injury, sternoclavicular joint separation (caused by a heavy blow to the chest), and bursitis (Christensen, 2002; Cisternino, 1978, p. 951; Fialka, 2004, p. 20; Hannon, 2006, p. 239; Jones, 2010, p. 66, 74; Quillen, 2004, p. 1947, 1948, and 1951; Stanley, 1988, p. 461).

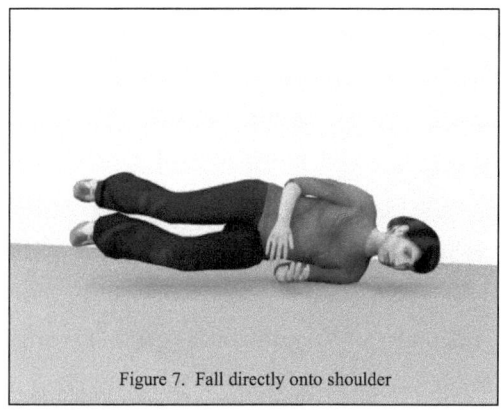

Figure 7. Fall directly onto shoulder

A fall directly onto the shoulder can cause a ligamentous tear, resulting in AC joint sprain/separation, brachial plexus injury, proximal humerus fracture, and clavicle fracture (Figure 7) (Christensen, 2002; Hannon, 2006, p. 241; Jones, 2010, p.66-68; Stanley, 1988, p. 461).

Traction injuries, a blow that forcibly tilts the head and neck to the side and depresses the shoulder, can result in subluxation or a brachial plexus injury. Swinging the arm strenuously into an awkward position can cause subluxation (Barnes, 1949, p. 10-11; Christensen, 2002; Jones, 2010, p. 74).

Breaking one's fall with a hand causes induced forces that are transmitted up the arm to the shoulder (Figure 8). This mechanism of injury causes clavicle fracture, posterior dislocation, and proximal humerus fracture (Cisterno, 1978, p. 951; Hannon, 2006, p. 241; Stanley, 1988, p. 461).

Figure 8. Break fall with hand

Overuse or repetitive motions (commonly repetitive overhead motions, often seen in sports injuries) can result in rotator cuff tears or tendinopathy, glenoid labrum tears, biceps brachii tendon tears, bursitis, and impingement (Bigliani, 1997, p. 1855; Jones, 2010, p. 70-72; McLeod, 1986, p. 1904; Quillen, 2004, p. 1953).

Degenerative and anatomical conditions within the shoulder are common causes of shoulder injury. Osteoarthritis is a degenerative joint disease caused by changes that are characterized by the abrasive wearing away of the articular cartilage concurrent with the reshaping of the adjacent ends of bones. As a result, masses of new bony protrusions, or osteophytes, occur. Impingement can be caused by osteophyte formation on the undersurface of a degenerative acromioclavicular joint. Impingement can also be caused by degenerative tendinopathy, which leads to partial rotator cuff tears and allows proximal migration of the humeral head, resulting in impingement and, ultimately, complete tears of the rotator cuff (Bigliani, 1997, p. 1855). Another factor leading to the development of impingement is the anatomical configuration of the acromion process, as described above.

## Associated Injuries

An injury to the shoulder is often complex, frequently resulting in associated injuries, and may even give rise to further injuries in the future (also summarized in Table #3). Clavicle fractures can lead to nerve and blood vessel damage, risk of osteoarthritis, and shoulder deformity (Jones, 2010, p. 66). Rotator cuff tears/injuries can lead to the development of bone spurs in the subacromial space (Jones, 2010, p. 70). Furthermore, between 14 to 63 percent of anterior dislocations are also associated with rotator cuff tears (Cutts, 2009, p. 4). Bursitis of the subacromial bursa is often associated with impingement syndrome (Hannon, 2006, p. 239). Impingement can also be associated with strain or loss of competency of the rotator cuff tendons, glenohumeral instability, calcification of the coracoacromial ligament, and rotator

46

cuff tears (Bigliani, 1997, p. 1857; Fongemie, 1998, p.667; Jones, 2010, p. 72). Anterior shoulder dislocation can be associated with proximal humeral fracture/fracture of the proximal head of the humerus, rotator cuff tears, and glenoid labrum tears (Cutts, 2009, p. 3-4; Hannon, 2006, p. 241). Posterior shoulder dislocation can be associated with humeral head fractures and avulsion fractures of lesser tuberosity (Cisternino, 1978, p. 951; Hannon, 2006, p.241). SLAP tears are associated with both anterior and posterior shoulder dislocations, as well as with rotator cuff tears (Mohana-Borges, 2003, p. 1450).

### Table #3. Commonly Associated Injuries of the Shoulder

| Shoulder Injury | Associated injuries |
|---|---|
| Clavicle fracture | Nerve and blood vessel damage, osteoarthritis, shoulder deformity |
| Rotator cuff tear | Bone spur in the subacromial space, anterior dislocation, impingement |
| Impingement | Bursitis, strain or loss of competency of rotator cuff tendons, glenohumeral instability, calcification of the coracoacromial ligament, rotator cuff tear |
| Anterior dislocation | Proximal fracture/fracture of proximal head of humerus, rotator cuff tear, glenoid labrum tear |
| Posterior dislocation | Compression fracture of humeral head, avulsion fracture of the lesser tuberosity |
| SLAP tear | Anterior dislocation, posterior dislocation, rotator cuff tear |

# Integrative Medicine Therapies for Shoulder Injuries

Integrative medicine (also referred to as complementary and alternative medicine, or CAM) treats the whole patient, not just the disease, by treating the mind, body, and spirit. Every patient is unique; therefore, a thorough understanding of the individual patient is essential and can be achieved through diet journals, patient interviews, and lab testing. Most integrative medicine programs combine conventional Western medicine (including imaging techniques) with alternative, or complementary, treatments and therapy, including herbal medicine, acupuncture, prolotherapy, massage, biofeedback, yoga, and stress reduction techniques.

## Acupuncture

Acupuncture aims to heal through the stimulation of anatomical points on the body, using a variety of techniques. Acupuncture involves penetrating the skin with thin, solid, metallic needles manipulated by the hands or by electrical stimulation. According to traditional Chinese medicine, acupuncture regulates the flow of qi, or vital energy, through the body, thus keeping the body in a balanced state (National Center for Complementary and Alternative Medicine, 2011). Reaves and Bong (2011) recommend four steps to acupuncture treatment of the injured/separated AC joint. Step one involves using points and techniques that may immediately decrease pain or increase range of motion. Step two involves using meridian and microsystem points outside of the injury site. Sep three uses points that benefit the qi, blood, and internal (zangfu) organs. Step four uses local and adjacent points at the site of the injury (p. 22-24).

## Homeopathy

Homeopathy utilizes remedies that stimulate self-healing. Homeopathy typically works under the principle of similars or "like cures like," in which a disease can be cured by a substance that produces similar symptoms in healthy people. Another important principle of homeopathic treatments is dilution: the lower the dose of the medication, the greater its effectiveness. Most homeopathic remedies are diluted so that no molecules of the substance remain; even so, the "essence" of the healing substance remains and cures the disease. Homeopathic remedies are individualized to each patient based on history, body type, and symptoms, and are derived from natural substances that come from plants, minerals, or animals (National Center for Complementary and Alternative Medicine, 2010). Barkauskas (2007) suggests Traumeel, Kalmia compositum, Ferrum-Homaccord, or Lymphomyosot for injuries to the ligaments, tendons and muscles of the shoulder, and suggests Silicea-Injeel and Thyreoidea compositum for chronic weakness of connective tissue of the shoulder (p. 13).

## Prolotherapy

Prolotherapy (proliferative therapy) involves a series of injections of irritants, osmotic shock agents, and/or chemotactic agents designed to stimulate inflammation in injured tissues, which leads to tissue repair and/or growth. Prolotherapy is used to treat many chronic shoulder injuries, including rotator cuff tears, arthritis, sprains, and AC joint separation (Van Pelt, 2009). When tissues are injured, inflammation stimulates substances carried in the blood, which produce growth factors in the injured area to promote healing. Ligaments, tendons, and cartilage, however, have poor blood supply and take longer to heal than other tissues; as a result, incomplete healing of these structures is common. Traditional treatments for ligament and tendon injuries include anti-inflammatory medications, nonsteroidal anti-inflammatory drugs (NSAIDS), or corticosteroids to temporarily relieve pain and/or swelling. Proponents of prolotherapy argue that by suppressing inflammation and, therefore, fibroblast proliferation and collagen formation, these traditional treatments actually suppress the body's natural healing process, and the injured tissues do not fully heal. As a result, many patients suffer from chronic ankle sprains, laxity, or instability due to incomplete healing (Alderman, 2007, p.11).

The injection of proliferants triggers a healing cascade, which begins with granulocyte infiltration (which brings newly formed blood cells, fibroblasts, and inflammatory cells), continues with monocyte/macrophage invasion (which destroys or neutralizes the injurious agent—in this case, the proliferant), and ultimately leads to the activation of fibroblasts and the formation of collagen (the major component of connective tissue, i.e. ligaments and tendons). Common irritants include phenol, quaicol, tannic acid, and quinine; these substances create a local tissue reaction, which causes granulocyte infiltration. Osmotic shock agents, such as glucose, hypertonic dextrose, glycerin, and zinc sulfate also create a local tissue reaction to stimulate granulocyte infiltration. Chemotactic agents such as sodium morrhuate cause a direct activation of local inflammatory cells (Alderman, 2007, p. 12; Schwartz, 1991, p. 221).

## Summary

A careful examination of the individual components of the shoulder joint helps to illuminate the complex anatomy of the entire shoulder. Similarly, through examination of the kinematics of the shoulder, the different types of mechanisms of injury become easier to understand. It becomes clear why the shoulder, with such a wide range of movement available to it, can be easily injured when its components exceed their physical limits through applied force. Finally, through the complementary techniques of Integrative Medicine, injuries to the shoulder can be treated in conjunction with treatment of the whole patient.

# References

Alderman, D. (2007, January/February). Prolotherapy for musculoskeletal pain. *Practical Pain Management*, 7 (1), 10-15. Retrieved from http://www.prolotherapy.com/ppm2007.pdf

Arnheim, D. D., & Prentice, W. E. (1993). *Principles of Athletic Training (8ᵗʰ ed.).* St. Louis: Mosby Year Book.

Barakuskas, D. (2007). A Biotherapeutic Approach to Common Sports Injuries. *Journal of Biomedical Therapy*, 1(1), 12-13.

Barnes, R. (1949). Traction Injuries of the Brachial Plexus in Adults. *The Journal of Bone and Joint Surgery*, 31 B (1), 10-16.

Bigliani, L.U., & Levine, W.N. (1997). Current Concepts Review Subacromial Impingement Syndrome. *The Journal of Bone and Joint Surgery*, 79 A (12), 1854-1868.

Christensen, K. (2002). Managing Shoulder Sprain/Strain Injuries. *Dynamic Chiropractic*, 20 (22). Retrieved from http://www.dynamicchiropractic.com/ mpacms/dc/article.php?id= 15417&no_paginate=true&p_friendly=true&no_b=true

Cisterno, S.J., Rogers, L.F., Stufflebam, B.C., & Kruglik, G.D. (1978). The Trough Line, A Radiographic Sign of Posterior Shoulder Dislocation. *American Journal of Roentgenology*, 130, 951-954.

Cutts, S., Prempeh, M., & Drew, S. (2009). Anterior shoulder dislocation. *Annal of the Royal College of Surgeons of England*, 91, 2-7. doi: 10.1308/003588409X359123.

Dox, I.G., Melloni, B.J., & Eisner, G.M. (1993). *The Harper Collins Illustrated Medical Dictionary*. New York: HarperCollins.

Fialka, C., Stampfl, P., Oberleitner, G., & Vécsei, V. (2004). Traumatic acromioclavicular joint separation—Current concepts. *European Surgery*, 36 (1), 20-24.

Fongemie, A.E., Buss, D.D., & Rolnick, S.J. (1998). Management of Shoulder Impingement Syndrome and Rotator Cuff Tears. *American Family Physician*, 57 (4), 667-674.

Hannon, P., & Knapp, K. (2006). *Forensic Biomechanics*. Tuscon: Lawyers & Judges, 237-241.

Huang, J.H., & Zager, E.L. (2004). Thoracic Outlet Syndrome. *Neurosurgery*, 55 (4), 897-903.

Jones, G., & Wilson, Ed. (2010). *Everyday Sports Injuries: Diagnosis, Treatment, and Prevention*. New York: DK Publishing, 66-74.

Kapit, W., & Elson, L. M. (2002). *The Anatomy Coloring Book (3rd ed.)*. San Francisco: Benjamin Cummings, 22-23, 31-32, & 54-56.

Martini, F.H., Bartholomew, E.F. (2000). *Essentials of Anatomy & Physiology, 2nd Ed.* Upper Saddle River, New Jersey: Prentice-Hall, 142-144, 151-153, 156, & 195-198.

McLeod, W.D., & Andrews, J.R. (1986). Mechanisms of Shoulder Injuries. *Physical Therapy*, 66 (12), 1901-1904.

McMinn, R.M.H., Hutchings, R.T., Pegington, J., Abrahams, P. (1996). *Color Atlas of Human Anatomy, 3rd Ed.* London: Mosby-Wolfe, 120 & 126.

Mohana-Borges, A.V.R., Chung, C.B., & Resnick, D. (2003). Superior Labral Anteroposterior Tear: Classification and Diagnosis on MRI and MR Arthrography. *American Journal of Roentgenology*, 181, 1449-1462.

National Center for Complementary and Alternative Medicine. (2010, August). Homeopathy: An Introduction. Retrieved from http://nccam.nih.gov/ health/homeopathy

National Center for Complementary and Alternative Medicine. (2011, August). Acupuncture: An Introduction. Retrieved from http://nccam.nih.gov/ health/acupuncture/introduction.htm

Patton, K.T., Thibodeau, G.A. (2000). *Mosby's Handbook of Anatomy & Physiology*. St. Louis: Mosby, 138-143, 164-167, & 202-207.

Quillen, D.M., Wuchner, M., & Hatch, R.L. (2004). Acute Shoulder Injuries. *American Family Physician*, 70 (10), 1947-1954.

Reaves, W., & Bong, C. (Summer 2011). Acupuncture Treatment of Shoulder Pain: The Acromioclavicular Joint. *The American Acupuncturist*, 56, 22-25.

Rogers, K. (2011). *Bone and Muscle: Structure, Force, and Motion*. New York: Britannica Educational Publishing, 78-83, 124-131, 138-141, 162-163, 218-221, & 230-233.

Schwartz, R.G., & Sagedy, N. (1991). Prolotherapy: A Literature Review and Retrospective Study. *Journal of Neurological and Orthopaedic Medicine and Surgery*, 12, 220-223.

Stanley, D., Trowbridge, E.A., & Norris, S.H. (1988). The Mechanism of Clavicular Fracture: A Clinical and Biomechanical Analysis. *British Journal of Bone and Joint Surgery*, 70-B, 461-464.

Woodward, T.W., & Best, T.M. (2000). The Painful Shoulder: Part 1. Clinical Evaluation. *American Family Physician*, 61(10), 2079-3088.

Van Pelt, R.S. (2009). Shoulder Prolotherapy Injection Technique. *Journal of Prolotherapy*, 1(4), 243-245.

Zuidema, G.D. (1997). *The Johns Hopkins Atlas of Human Functional Anatomy, Fourth Edition*. The Johns Hopkins University Press, 11 & 14.

# About the Authors

Kenneth Alvin Solomon, Ph.D., P.E., Post Ph.D., obtained a Bachelor of Science, Master of Science and Doctorate in Engineering, as well as a Post-doctorate in Risk Benefit Assessment from UCLA. Dr. Solomon also holds a Professional Engineering License. Dr. Solomon's studies are limited primarily to accident reconstruction, biomechanics, and risk-benefit assessment as demonstrated by his 39 years of independent research; his more than 200 internationally distributed publications, reports, and presentations; his thirteen book co-authorship; and his journal guest editorships. In December of 1998, and after nearly 23 years of service, he retired as Senior Scientist with the RAND Corporation. He was on the faculty at the RAND Graduate School for eighteen years, and has taught as an Adjunct Faculty at UCLA, USC, Naval Post-Graduate School, and George Mason University. Dr. Solomon has published studies in Transportation Accidents (automotive, trucks, motorcycles, bicycles); Industrial & Recreational Accidents (pressure vessels, rotating machinery, forklifts and cranes, exercise, gym, & recreational equipment, swimming pools, manufacturing and punch presses); Slip- or Trip-and-Fall Accidents; and Adequacy of Warnings.

Anne J. Yatco, B.S., M.F.A., obtained a Bachelor of Science degree in Biomedical Engineering with an emphasis in biomechanics from Marquette University in Milwaukee, Wisconsin, and a Masters in Fine Arts degree in Acting from the California Institute of the Arts in Valencia, California. Her studies included classical mechanics, CAD, physiology, and biochemistry. She also assisted in research projects through Marquette University's Biomedical Engineering Department. Ms. Yatco utilizes her knowledge of classical mechanics, biomechanics, and the mechanics of injury at the Institute of Risk & Safety Analyses to determine the potential for injury in a given accident.

Printed by Books on Demand GmbH, Norderstedt / Germany